THIS BOOK
BELONGS TO:

To my brother Jesse and mom, thank you for always believing in me. Dad, I hope I can be as strong as you. Thank you to my wonderful family and friends. Your support helped this dream come true.

Play Along Publishing
Reynoldsburg, OH

Illustrated by Leticia Ribeiro

Edited by Brooke Vitale
Book Design by Arlene Soto, Intricate Designs

Names: Dixon, Christina, author. | Ribeiro, Leticia, illustrator.
Title: I am the moon / by Christina Dixon; illustrations by Leticia Ribeiro.
Description: Reynoldsburg, OH: Play Along Publishing LLC, 2022. | Summary: A young dreamer wakes up as the moon. He experiences different space events, while being guided by the Man in the Moon.
Identifiers: LCCN: 2022917031 | ISBN: 979-8-9866488-1-1 (hardcover) | 979-8-9866488-3-5 (hardcover) | 979-8-9866488-2-8 (paperback) | 979-8-9866488-4-2 (paperback) | 979-8-9866488-0-4 (ebook)
Subjects: LCSH Space--Juvenile fiction. | Solar system--Juvenile fiction. | Dreams--Juvenile fiction. | Moon--Juvenile fiction. | Science fiction. | BISAC JUVENILE FICTION / Bedtime & Dreams | JUVENILE FICTION / Imagination & Play | JUVENILE FICTION / Science Fiction / Space Exploration
Classification: LCC PZ7.1 .D58 Ia 2023 | DDC [E]--dc23

I am the Moon

by Christina Dixon

Illustrations by Leticia Ribeiro

The Nebulan guardians zipped through the galaxy, searching.
They were seeking dreamers. But all they found was darkness.
Suddenly, a tiny light began to twinkle on Earth.

It was time.

"Daddy," Cornelius said, gazing at the moon, "who lives up there?
Do they turn the lights on at night and off during the day?
Do you think they get tired of eating cheese?"

"Cheese?" Dad said with a chuckle.
"Why would they eat cheese?"
"Because . . . the moon is made of cheese,"
Cornelius said. "Isn't it?"

Dad smiled and tucked the boy into bed.
"Those sound like questions for the
Man in the Moon. Maybe one day
you'll get to ask him."
"The Man in the Moon,"
Cornelius whispered as
his eyes started to close.
"The Man in the . . ."

High above the Earth, a fisherman's line
swayed back and forth . . . back and forth.
The line whistled through the clouds,
heading straight toward Cornelius's window.

"Gotcha," said a soft, deep voice.

Cornelius turned over as the line was being reeled in.
The night breeze stung his skin, and his fingers
reached for his blanket, but there was nothing there.

He flicked his groggy eyes open to find where his
blanket had gone . . . and found himself facing
Saturn's huge ring!

Cornelius blinked again, now wide awake.
Looking around, he saw Earth, the sun,
Jupiter, Venus, rainbows, and thousands
of stars all around him.

"Am I in **space!?**" he asked.

"Yes," a warm, soft voice replied.

Cornelius gulped.
"Who said that?
Where are you?"

"I am the Guardian of the Moon," the voice said.
"Though most people call me the Man in the Moon."
"The Man in the Moon!" Cornelius exploded.
"But . . . why am I here?"

The Guardian laughed.

"You are here because it's your time,
Cornelius. You are a dreamer.
You see how the lights on Earth go dark?
That is the cloud of doubt. The only force
stronger than doubt is your light.
It shines on the Earth, spreading hope.
You are here because you are special.
Your spark of wonder makes you shine."

For a moment, the boy made no sound.
Then he shouted,
"I am the moooooon!!"
His eyes swept the space around him.
"But wait, there are rainbows in space?"
"Oh, yes, my boy. The rainbows come from the sun's
rays bouncing off Earth during a full Earth.
This is called Earthshine."

"Earthshine," Cornelius whispered.
Then, "Are you from Earth?"
"Guardians come from the dust that forms when a star is born.
We are called Nebulans," the Guardian proudly replied.
"Those darts of light you see are my family."
Cornelius couldn't believe what he was seeing.
"I always thought those were shooting stars!"

Sweet music filled Cornelius's ears.
"What is that?" he asked.
"Ah," said the Guardian. "That is the
sound of a star vibrating.
It is how we communicate up here."
"Communicate? Can I talk to my dad?"
The Guardian shook his head.
"I'm sorry. But know this: you are safe.
Time works differently here.
Your father will never even know
you are gone."

The Guardian checked his watch.
A deep rumbling sound now
covered the music. It seemed
distant, but it was getting closer.
Something was heading their way.

"Ah, here we are," said the Guardian. "5...4...3..."
"Why are you counting?" Cornelius asked. "What is—
Ha HA HAAAA! Stop tickling me!" the boy shouted.

"III'mmmm nnnnot tickkkling you," the Guardian's voice vibrated.

"Yoooou are feeling a moonquake."

The shaking went on for several minutes.
Then, just as quickly as it had started, it
stopped. The moon made a low growl.
"Oh," said the Guardian, "you must be
hungry. Never fear, food is near."

Cornelius followed the Guardian's gaze toward a group of rocks heading their way. "But those are **meteors**," Cornelius said. "Don't you eat the cheese from the moon?"

The Guardian let out a roaring laugh. "Nooo, that myth came from another dreamer."
The boy's face lit up with excitement. "There are more moon kids?"
"Of course, my boy." The Guardian leaned back to look at the stars. "Dreamers can be anything. That is why you are so special."

"Watch this," the Guardian said.
Closing his eyes, he stood still, like a rocket ready to launch.
His body shook, and he began to glow brighter and brighter.
Then . . . **BOOM!** He zipped through the sky, swallowing
every rock in his path with a net.

The rocks glowed and buzzed as the Guardian deposited them on the moon's surface. "Go on. Try one!"
"A rock?" Cornelius asked. "Thanks, but . . . I don't think I am that hungry anymore."
The Guardian smiled.
"My dear dreamer, things are not always as they seem on Earth."

The rock fizzed and popped as he bit into it. Suddenly, it exploded with flavor. All different types of food flooded his mouth.

Pizza, cotton candy, cheesy fries, glazed donuts, hot dogs. Mmmmm . . . RAVIOLI!

"Is . . ." Cornelius began, breathing heavily. "Is it . . .
hot to you? Is it daytime? It's REALLY bright!"
The Guardian turned to find the moon no more than
a soft shadow. "Cornelius, where have you gone?"

The boy was confused. "I'm right here."
"Oh, yes! That's right!" said the Guardian.
"Time has moved so fast. You are going through a new moon phase.
You are between Earth and the sun. That is why you feel hot and it
looks too bright. The sun is directly behind you!"

"So," Cornelius said,
"there's no light bulb in the moon?"
"Nope," replied the Guardian.
"All of your light comes from the sun."

"You have gone through the entire moon phase," the Guardian continued, "which means it is time to go home." Sadness filled Cornelius's heart. "But—" he began. Reaching out, the Guardian took the boy's hand. "We have one more adventure," he whispered. "Shall we travel by light or—"

"BY LIGHT!!" the boy shouted.

The two stood still, like
rockets ready to launch.
Their bodies shook, and they grew brighter
and brighter. At their brightest, they zipped past the
rainbow and back to Earth. Looking back, Cornelius saw
the planets get
smaller
and
smaller.

His head rested softly on his pillow.
He heard a whisper say, "Never stop dreaming."

Cornelius slowly opened his eyes to see his dad smiling at him.
"**Dad!**" He leaped into his dad's arms.
"Good morning," Dad said.
"You're hugging me like you haven't seen me in a month. What on Earth did you dream about last night?"

Cornelius grinned and told his dad about his adventure as the moon.

High above the Earth, the Moon Guardian
watched Cornelius. He was shining like a star.
As the boy grew older, his light would continue
to shine. And that light would spread, until
the Earth was covered in lights shining as
brightly as his own.

About the Author
Christina Dixon

I graduated from Wilberforce University with a Bachelor of Science degree. While growing up, I had a vivid imagination and a healthy dose of curiosity. As a child, I wrote stories about kids engaging in accidental science experiments and robust adventures. I was inspired by movies and shows such as *Honey, I Shrunk the Kids*, *Reading Rainbow*, and *The Magic School Bus*.

My creative side was fired up during a bedtime story when my four-year-old son, Jabari, asked if the moon was made of cheese. We exchanged stories about who would eat the cheese and how they got there! That night, the gates to my imagination opened, and I never want to close them again. My story intentionally uses science terms like *nebula* and *earthshine* to encourage further learning, with a theme of never stop dreaming because dreamers are special. I hope to continue to create more stories that will encourage adventure and ignite the imagination.

About the Illustrator
Leticia Ribeiro

Leticia is a Brazilian native from Sao Paulo. At a young age, she knew that drawing was her passion. She graduated from Anhembi Morumbi University with a degree in architecture and had great success with interior designing, specifically building schematics. Her career shifted to illustration after she moved to Ireland, when a graphic artist noticed her work.

After landing her first freelance job, she decided to make illustration her career. Today she finds inspiration in Paola Escobar's work to illustrate projects such as *As Cores dos Pensamentos de Tina* by Marismar Borém. Recently, she was nominated as one of the top ten finalists for the Penguin Cover Design Award. Leticia continues to work on multiple projects to sharpen her craft.

For fun activities
and glossary terms
from the book,

scan the QR code